UDDERWISE

By the same author:
Pigmania

UDDERWISE
EMIL VAN BEEST

FARMING PRESS
Wharfedale Road, Ipswich, Suffolk IP1 4LG

British Library Cataloguing in Publication Data

Beest, Emil van
    Udderwise.
    1. Dutch wit and humor, Pictorial
    I. Title
    741.5′9492   NC1549

    ISBN 0-85236-176-9

First published 1988

ISBN 0 85236 176 9

Cover design: Hannah Berridge

Printed and bound in Great Britain by
Biddles Ltd, Guildford and King's Lynn

Emil van Beest comes from Goor, in the Netherlands. His cartoons have been a regular feature of the Dutch agricultural magazine, *Boerderij*, since 1965 and have appeared throughout the world.

A collection of van Beest's cartoons about pigs, *Pigmania*, has also been published by Farming Press.

FARMING PRESS

Farming Press publishes a wide list of books about farming.

On the humorous side there are books by Peter Ashley, Henry Brewis and John Terry, each of whom sees the funny side of agriculture in his own distinctive way.

The range of practical books for farmers and students published by Farming Press is unrivalled in Britain. The list includes titles on pig, sheep, dairy and arable farming as well as many other farming and veterinary topics. Farming Press also publishes three monthly magazines – *Arable Farming*, *Dairy Farmer* and *Pig Farming*.

For a free illustrated catalogue of books or specimen magazines please contact:

Farming Press, Wharfedale Road, Ipswich IP1 4LG.